Preparing for
CALCULUS

The Language, Concepts, and Skills Needed for Success

Jack McCabe

Copyright © 2024 by Jack McCabe

ISBN: 978-1-77883-436-3 (Paperback)

All rights reserved. No part of this publication may be reproduced, distributed, or transmitted in any form or by any means, including photocopying, recording, or other electronic or mechanical methods, without the prior written permission of the publisher, except in the case brief quotations embodied in critical reviews and other noncommercial uses permitted by copyright law.

The views expressed in this book are solely those of the author and do not necessarily reflect the views of the publisher, and the publisher hereby disclaims any responsibility for them. Some names and identifying details in this book have been changed to protect the privacy of individuals.

BookSide Press
878-7408,090
www.booksidepress.com
orders@booksidepress.com

Contents

Preface v
I. The Language Of Functions 1
II. Polynomials 3
III. Lines, Circles And Intervals 11
IV: Exponents Explained 15
V. Radicals And Rational Exponents 18
VI. Exponential Functions 21
VII. Inverse Functions 25
VIII. Inverse Of Exponential Functions 27
IX. Exponential And Logarithmic Graphs 32
X. Rational Algebraic Functions 35
XI. Trigonometric Functions 42
Appendix A
    The Number Zero 48
Appendix B
    Dealing With Numerical Fractions 50
Appendix C
    Euler's e 52
Appendix D
    Answers Practice Problems 54

PREFACE

Why do students find calculus so challenging? Success with calculus problems requires the ability to readily identify the mathematics needed to solve a given problem – an attribute called **mathematical fluency**. Experience has taught me that only the most accomplished math students enter a beginning course in calculus with sufficient mathematical fluency. Let me explain further.

In high school, students are often reminded of a formula from middle school: Distance equals rate times time. But in calculus, rates are most often changing. In fact, problems are given where the rate of several dimensions are changing at different rates. For example, consider water flowing into an inverted right circular cone (vertex is pointed downward like a paper drinking cup). Suppose the water is flowing into the inverted cone at a constant rate of 5 cubic inches per second, then the depth of the water is rising at decreasing rate, the diameter of the top circle of water will be increasing at a deceasing rate and the surface of the cone will be getting wet at a slowly increasing rate. These three rates will be related by the geometry of the cone.

For another example from calculus, consider the problem: *"A long rope is attached to the bow of a small boat located 150 feet from a dock that is 20 feet above the water. From the top of the dock, the rope is pulled in at a steady rate of 10 feet per minute. How fast is the boat approaching the dock when the boat is 80 feet from the dock?"*

The length of the rope is getting shorter at a constant rate while the boat is approaching the dock at an increasing rate. To solve this problem, students will draw a triangle where the dock is perpendicular to the water and the rope is the hypotenuse of a right triangle. Is this problem a case of similar triangles or a case of the Pythagorean Theorem?

These two examples illustrate how calculus application problems require both a broad and deep knowledge of high school math. But there are reasons to be encouraged. Students now use graphing calculators to solve equations. Additionally, some of the math studied in high school is no longer needed for the first two semesters of calculus. This includes complex numbers, matrices and determinants as well as probability and statistics. I wrote this book to specify the math you will need to know for calculus and to provide some related practice problems. Complete solutions are provided in an appendix. It is my hope that you will find the material informative and an effective resource for preparing you for your first experience with calculus.

Before my recent retirement from the classroom, I spent more than forty-five years teaching mathematics – preparing high school students for college. I spent a few years teaching at the college level and over twenty-five years teaching calculus, including Advanced Placement Calculus. Many of my students needed individual help. I enjoyed helping them and I learned much from these one-on-one sessions. Helping those students inspired me to write this book.

. . .

THE LANGUAGE OF FUNCTIONS

Perhaps you are already aware that an equation such as $y = 3x^2 - 5x + 4$ is considered a function of x. This relation is also expressed as $f(x) = 3x^2 - 5x + 4$. Most problems in calculus are concerned with functions. Here is a sample of other types of functions you have most likely seen:

$$f(x) = \frac{2x-1}{3x^2-5} \qquad f(x) = \sqrt{4x^2+1} \qquad f(x) = 2e^{.5x}$$

$$f(x) = \ln|x+1| \qquad f(x) = 2\sin(x)\tan(x) \qquad f(x) = x^{\frac{2}{3}}$$

In turn, I will discuss each type, but for now I review the vocabulary and terms associated with the **vocabulary** used when functions are discussed. When a **formula** contains two variables, where the values for one variable y **depend arithmetically** or **algebraically** on values of a second variable x, we say **y is a function of x** and denote this dependence with the **notation** $y = F(x)$. The formula provides a **rule of the function.**

By the **domain** of a function we mean the values of the **independent variable** (usually x) allowed by the problem. Sometimes the domain of a function has restrictions related to arithmetic such as division by zero (neither $\frac{5}{0}$ nor $\frac{0}{0}$ have a numerical value). And the square-roots of a negative number (e.g. $\sqrt{-4}$) has no value. Other times the domain is restricted by the physical properties of the objects being discussed

(e.g. dimensions such as length, area and volume are never negative.)

The set of values of the **dependent variable** (usually y) associated with domain values is called the **range** of the function. The **zeros** of a function are the domain values for which the range value equals zero. On a graph of a function, the zeros are the x-axis intercepts.

. . .

Please see the **Appendix A** for two useful observations regarding solving equations. See **Appendix B** for a discussion of why dividing by zero is disallowed in arithmetic and algebra.

. . .

The discussion of each types of function - polynomial, rational algebraic, exponential, logarithmic, trigonometric - will begin with one or more examples of the function. The discussion will use the language associated with functions.

. . .

POLYNOMIALS

In calculus, students will encounter **polynomial functions,** the type least used in science and engineering. Students usually find it easy to understand and use polynomial functions.

Calculus students will encounter problems similar to the following: *A rock is thrown almost straight upward off of the roof of a building with a velocity of 60 mph. The rock will first travel upward, then fall downward to the ground. If the rock leaves the thrower's hand at 48 feet above the ground, a mathematical model for this situation is the formula $H = -16t^2 + 88t + 48$.*

The variable H represents the height of the rock in feet above the ground. The variable t represents the time in seconds. The 48 in the formula is the initial height in feet of the rock. The 88 is the initial speed in feet per second. It is the result of converting 60 miles per hour to feet per second. The -16 (feet per second per second) is the effect of gravity forcing the rock downward. How much time will it takes for the rock to hit the ground? What is the maximum height reached by the rock?

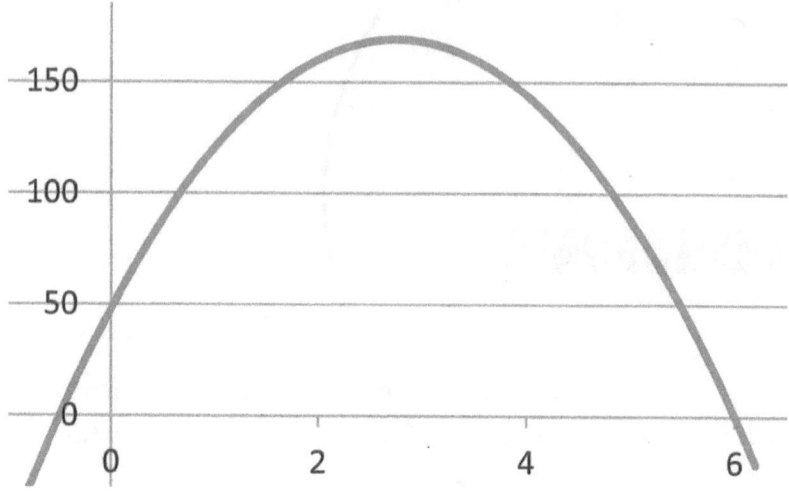

Using a graphing calculator and the graph of the polynomial function, $y = -16x^2 + 88x + 48$ we can determine the zeros as $x = \frac{-1}{2}$ and $x = 6$. The rock hits the ground exactly 6 seconds after it is thrown. In this situation, time is positive, so the solution $x = \frac{-1}{2}$ is irrelevant to the time of the rock's travel. Because this graph has **symmetry**, the maximum height occurs at $x = \frac{11}{4}$, the midpoint between $x = \frac{-1}{2}$ and $x = 6$.

The maximum height reached is $y\left(\frac{11}{4}\right) = 169\ feet$. The **domain** of this function is $0 < x < 6$ with a corresponding **range** $0 < y < 169$.

The rule of the function is the expression $-16x^2 + 88x + 48$, an example of a **second degree** polynomial, also called a **quadratic** expression. The graph of a second degree polynomial function is called a **parabola**. The **standard form** of a parabola is $y = ax^2 + bx + c$ and the **vertex form** is $y = a(x - h)^2 + k$, where (h, k) are the coordinates of the vertex.

The vertex of the graph of $y = -16x^2 + 88x + 48$ is the maximum point $(\frac{11}{4}, 169)$, so the vertex form of the equation is $y = -16(x - \frac{11}{4})^2 + 169$.

Polynomials

Further on we will be able to graph the actual path of objects moving under the influence of gravity given its initial angle of ascent and initial velocity e.g. baseballs, footballs, basketballs, etc.

Consider a second problem: *A commercial refrigeration company will be building cooling units, each requiring a pan to collect excess water. The pan will be installed six inches below the cooling unit. The company proposes to manufacture the pans by cutting squares from each corner of a rectangular sheet of metal measuring 30 inches by 40 inches. The excess metal on each side is folded vertically then welded (see figure below). 'The company requires the pan to hold as much water as possible, so each pan constructed from the 30 inch by 40 inch metal sheet must have a volume as large as possible. Determine the length of the sides of the squares that provide the maximum volume.*

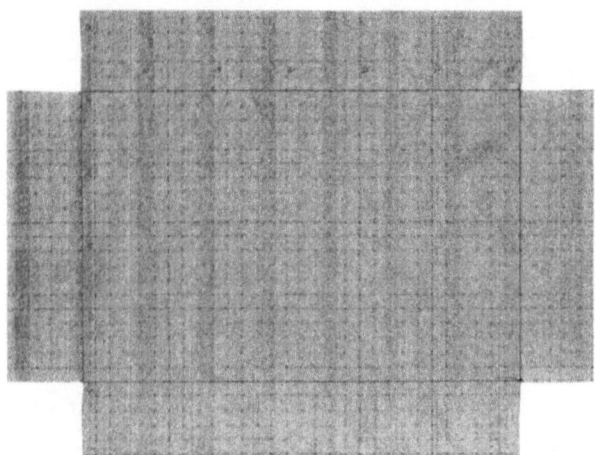

To understand the problem, first calculate the volume when the four squares are each 5 inches by 5 inches. The short side of the base of the pan will be $30 - 2 * 5 = 20$ inches, because 5 inches are required for the two squares on each 30 inch side.

The long side of the base of the pan will be $40 - 2 * 5 = 30$ inches because 5 inches are required for the two squares on each 40 inch side. The depth of the pan will be 5 inches. The volume of the pan will be 3000 cubic inches (20 * 30 * 5 = 3000).

These calculations suggest using the polynomial function $V(x) = x(30 - 2x)(40 - 2x)$ to explore this problem. V represents the volume and x represents the length of a side of a square. Use a graphing calculator to view the graph of the equation $y = x(30 - 2x)(40 - 2x)$.

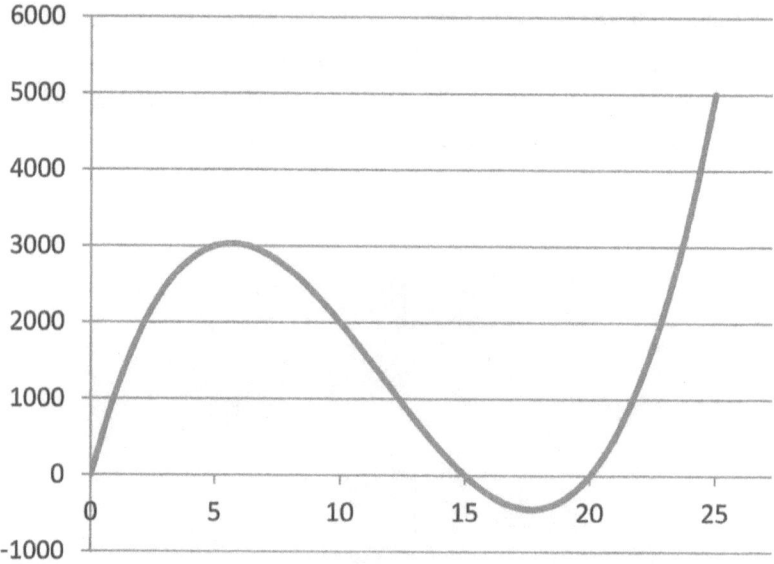

Entering $y(5)$ yields 3000. This confirms that our formula is consistent with the previous calculations. Continue the experiment with $x = 6$ and $x = 7$.

You should have $y(6) = 3024$ and $y(7) = 2912$. The volume is less for $x = 7$ than for $x = 6$ so experiment with $x = 6.5$. You should have $y(6.5) = 2983.5$, a volume still less than $y(6)$, so try $y(5.5) = 3030.5$.

Exploring the graph of $y = x(30 - 2x)(40 - 2x)$ with a graphing calculator will enable you to determine that the maximum volume of the pan.

You should see that the maximum volume is approximately 3032.3 cubic inches (52.5 quarts) when x equals approximately 5.66 inches.

For our function $y = x(30 - 2x)(40 - 2x)$ the values of x are restricted to $0 < x < 15$, the **domain** of our function. The **range** of our function is $0 < y < 3032$.

The **zeros** of the function $y = x(30 - 2x)(40 - 2x)$ are $x = 0$, $x = 15$ and $x = 20$. The formula $y = x(30 - 2x)(40 - 2x)$ provides a **rule** for this third degree polynomial function. Its expanded rule is $4x^3 - 140x^2 + 1200x$.

The standard form of the rule of a **third degree** polynomial function is $ax^3 + bx^2 + cx + d$.

The standard form of the rule of an nth degree polynomial is
$a_n x^n + a_{n-1} x^{n-1} + a_{n-2} x^{n-2} + \cdots a_2 x^2 + a_1 x + a_0$.

The letters multiplying the powers of x are constants, called **coefficients**.

An nth-degree polynomial function will have (n +1) terms, where some of the coefficients might be zero. The final term which is a constant can be considered to be multiplying the constant times x^0.

A graphing calculator can reveal the zeros, the **local maximums** and **local minimums** as well as the zeros of a polynomial function. For example, consider the graph of the polynomial $y = 2x^4 + 5x^3 - 59x^2 - 92x + 60$.

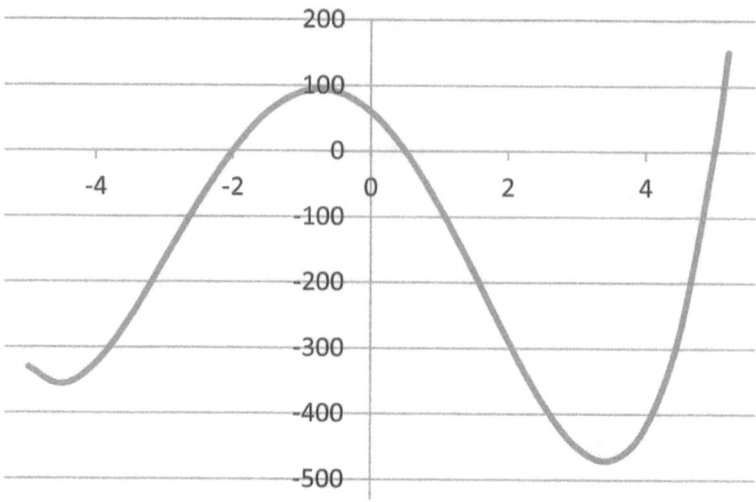

A graphing calculator will reveal a relative maximum at $(-.74, 94.35)$ and relative minimums at $(-4.56, -356.65)$ and $(3.42, -471.11)$. The calculator will also reveal zeros (x-axis intercepts) at $x = -6$, $x = -2$, $x = \frac{1}{2}$ and $x = 5$. These intercepts provide the linear factors. The polynomial can be factored: $(x + 6)(x + 2)(2x - 1)(x - 5)$.

In general, the standard form of the rule of a polynomial function can be expressed as a product of linear factors and prime quadratic factors. Each factor will have real coefficients (as opposed to complex coefficients.)

. . .

Practice Problems: See solutions in Appendix D

Use a graphing calculator to express each polynomial as a product of linear and quadratic factors.

1. $6x^5 + 19x^4 - 5x^3 - 25x^2 - x + 6$

2. $3x^4 + x^3 - 4x^2 - 7x - 5$

Polynomials

Three hundred books currently sell for $40 each, resulting in a revenue of (300)($40) = $12,000. Research indicates that for each $5 increase in the price, 25 fewer books are sold.

3. Write the revenue R as a function of the number x of $5 increases.

4. What selling price will bring the maximum revenue?

A 24 inch by 36 inch sheet of cardboard is folded in half to form a 24 inch by 18 inch rectangle as shown in the **figure below**. Four congruent squares of side length x are cut from the corners of the folded rectangle. The sheet is unfolded and the six tabs are folded up to form a cardboard suitcase.

5. Write a formula $V(x)$ for the volume of the suitcase.

6. Find the domain of V.

7. Find a value of x that yields a volume of 1120 cubic inches.

8. Use a graphing calculator to find the maximum volume and the value of x yielding this volume.

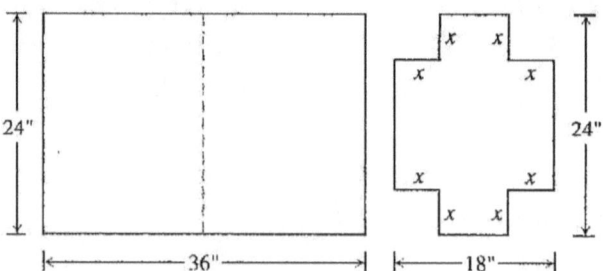

The sheet is then unfolded.

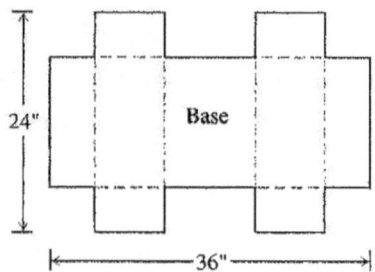

A rock is thrown upward by an astronaut standing on the moon with a velocity of 100 mph. If the rock leaves the astronaut's hand at 6feet above the surface of the moon, the mathematical model for this situation is $H = -2.65t^2 + 146.67t + 6$.

The variable H represents the height of the rock in feet above the surface of the moon. The variable t represents the time in seconds. The 6 in the formula is the 6feet from the initial height of the rock. The -2.65 (feet per second per second) is the effect of gravity on the moon, forcing the rock downward.

9. How much time will it takes for the rock to hit the ground?

10. What is the maximum height reached by the rock?

LINES, CIRCLES AND INTERVALS

The graphs of the simplest polynomials are lines. The graph of the equation $y = 3$ is a horizontal line, while the graph of the equation $x = 5$ is a vertical line.

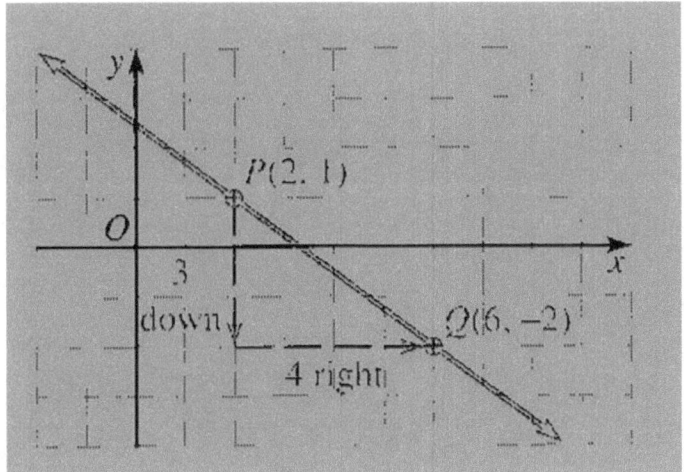

The graph of the line shown above reveals the slope as $\dfrac{-2-1}{6-2} = \dfrac{-3}{4}$.

Using the point (2, 1), a **point-slope form** of the line is $y = \dfrac{-3}{4}(x - 2) + 1$.

Any point on the line can be used to create a **point-slope form** of the line. The point (6, -2) is also on the line and provides another equivalent equation $y = \dfrac{-3}{4}(x - 6) - 2$.

11

The **slope-intercept form** of the equation of the line is $y = \frac{-3}{4}x + \frac{5}{2}$.

The **standard form** of the equation of the line is $3x + 4y - 10 = 0$.

. . .

The center-radius form of circle shown is $(x - 2)^2 + (y + 3)^2 = 36$. The graph is not the graph of a function because there are x-values with two different corresponding y-values (e.g. both $(2,3)$ and $(2,-9)$ are on the circle).

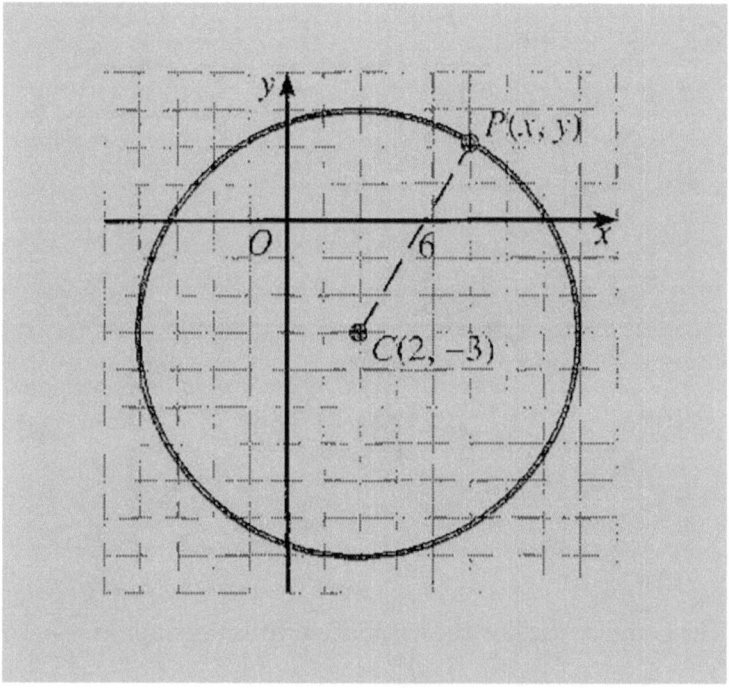

Graphs of circles do not satisfy the vertical line test.

The graph of the top half of our circle is shown below. It has the equation $y = -3 + \sqrt{32 + 4x - x^2}$ with domain $-4 \leq x \leq 8$ and range $-3 \leq y \leq 3$.

Lines, Circles And Intervals

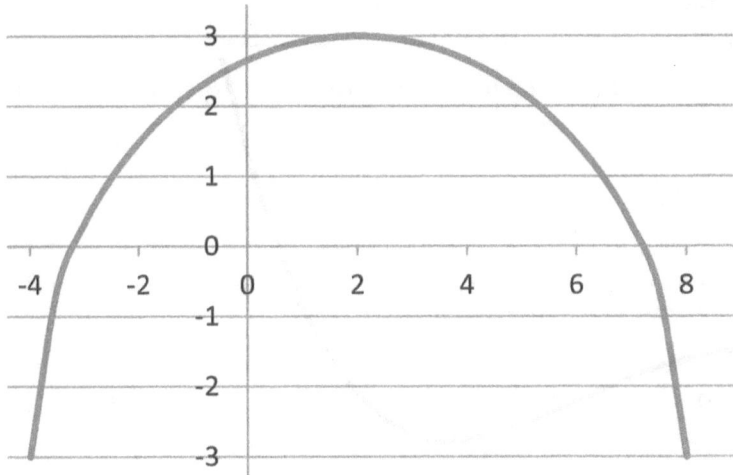

The distance between two numbers is the **absolute value of their difference**. For example the distance between -7 and 12 is $|-7 - 12| = |12 - (-7)| = 19$. In general, the distance between x and y is $|x - y|$. The equation $|x - 12| = 19$ states the distance between x and 12 equals 19. We calculate 19 units to the right of 12 (12 + 19 = 31) and 19 units to left of 12 is (12 − 19 = −7). So $x = 31$ and $x = -7$ are the two solutions to the equation $|x - 12| = 19$.

Solutions to inequalities are often **intervals of numbers on the x-axis**. For example, to solve the inequality $|4x + 7| < 12$, divide both sides by 4 to produce the equivalent inequality $|x - \frac{-7}{4}| < 3$. This inequality states the distance between x and $\frac{-7}{4}$ is less than 3. So the solutions are numbers between $\frac{-7}{4} - 3$ and $\frac{-7}{4} + 3$. The is the **open interval** $\frac{-19}{4} < x < \frac{5}{4}$ also expressed as $(\frac{-19}{4}, \frac{5}{4})$.

Practice Problems:

Express the solution to each inequality as an interval or a combination of intervals:

11. $\left|\frac{2}{3}x - \frac{4}{5}7\right| \leq 6$

12. $|2x + 3| \geq 1$

EXPONENTS EXPLAINED

Consider the problem: *The radius of each of the six electrons in a carbon atom is approximately 2.8E-13 meters and the radius of the carbon atom is 7.0E-9 meters. Find the proportion of space in the atom contained by the electrons.*

Let's review **exponents** and their **properties**. The notation 2^5 symbolizes the number 2 raised to the power 5, and is an abbreviation for $2*2*2*2*2$. In the expression of 2^5 the number 2 is called the **base** and the power 5 is called the **exponent**.

The base B of an exponential expression B^M is restricted to numbers larger than zero. In most instances, the number 1 is disallowed as a base because 1 raised to any power is again 1, i.e. $1^x = 1$ for all values of x. The equation $1^x = 1$ has an infinite number of solutions while the equation $0^x = 1$ has no solutions (0^0 has no meaning).

Exponents have important properties, each consistent with common arithmetic. For example, $3^2 * 3^5$ is an abbreviation for $(3*3) * (3*3*3*3*3)$ which equals $3*3*3*3*3*3*3$ and is abbreviated by 3^7 which equals 3^{2+5}.

We have exponent property #1: **For $B > 0$, $B^M * B^N = B^{M+N}$**

Another property of exponents can be seen by considering $\frac{3^7}{3^2}$, an abbreviation for $\frac{3*3*3*3*3*3*3}{3*3}$ which equals $3*3*3*3*3$ and is abbreviated by 3^5. So $\frac{3^7}{3^2} = 3^{(7-2)}$.

We have exponent property #2: **For $B > 0$, $\frac{B^M}{B^N} = B^{M-N}$**

Another property of exponents can be seen by considering $\frac{3^5}{3^5}$ which by property 2 equals $3^{(5-5)}$ which equals 3^0.

Since $\frac{3^5}{3^5} = 1$, $3^0 = 1$

We have exponent property #3: **For $B > 0$, $B^0 = 1$**.

Consider 3^{-1} which can be expressed as $3^{(4-5)} = \frac{3^4}{3^5}$, an abbreviation for $\frac{3*3*3*3}{3*3*3*3*3}$ which equals $\frac{1}{3}$. So $3^{-1} = \frac{1}{3}$.

We have property #4: **For $B > 0$, $B^{-1} = \frac{1}{B}$.**

By extension we have exponent property #5: **For $B > 0$, $B^{-N} = \frac{1}{B^N}$.**

For another property of exponents, consider $(2*3)^4$, an abbreviation of $(2*3)*(2*3)*(2*3)*(2*3)$ or $(2*2*2*2)*(3*3*3*3)$. This is abbreviated by $2^4 * 3^4$.

So $(2*3)^4 = 2^4 * 3^4$ We have exponent property #6: **For $A > 0$ and $B > 0$, $(AB)^N =$**

The last exponent property we discuss comes from observing $(3^2)^4$, an abbreviation for $(3^2)(3^2)(3^2)(3^2)$ which equals $3^{(2+2+2+2)}$ or 3^8 by property #1. So $(3^2)^4 = 3^{2*4}$.

We have exponent property #7: **For $B > 0$, $(B^M)^N = B^{(M*N)}$**

Exponents Explained

. . .

We are now ready to answer the question posed earlier. The radius of each of the six electrons in a carbon atom is approximately $2.8E - 13$ meters and the radius of the carbon atom is $7.0E - 9$ meters. The letter E in the mathematical expression $2.8E - 13$ stands for the word **exponent base 10**. To find the proportion of space in the atom contained by the six electrons, we form the fraction: $\frac{6*2.8*10^{-13}}{7.0*10^{-9}}$ and use the properties of exponents to simply the ratio. The ratio equals $\frac{16.0*10^{-13}}{7.0*10^{-9}} = \frac{1.6*10^{-12}}{7.0*10^{-9}} = \frac{1.6}{7.0} * 10^{9-12} = \frac{1.6}{7000}$ or approximately $\frac{229}{10000}$.

The proportion of space in the atom contained by the electrons is approximately 0.00023. A carbon atom is mostly space.

. . .

Practice Problems:

Solve each equation for the variable N.

13. $(2^{3N})^2 = (2^N)^3 * 2^{N+1}$

14. $4^{N+3} * 16^N = 8^{3N}$

15. $\frac{2^{3N}}{2^N} * \frac{2^N}{2^{5N}} = \frac{4^{-N}}{4^{3N}}$

16. *The radius of each of the 79 electrons in a gold atom is approximately 2.8E-13 meters and the radius of the gold atom is $2.88E - 10$. Find the proportion of space in the gold atom contained by the electrons.*

RADICALS AND RATIONAL EXPRESSIONS

The properties of exponents will enable us to make a discovery. How might we interpret $9^{\frac{1}{2}}$? If we assume that the properties of exponents extend to fractional exponents, then by property #1 $(9^{\frac{1}{2}}) * (9^{\frac{1}{2}})$ equals $9^{(\frac{1}{2}+\frac{1}{2})}$, which equals 9.

How might we interpret $8^{\frac{1}{3}}$? If we assume that the properties of exponents extend to fractional exponents, then by property #1 $(8^{\frac{1}{3}}) * (8^{\frac{1}{3}}) * (8^{\frac{1}{3}})$ equals $8^{(\frac{1}{3}+\frac{1}{3}+\frac{1}{3})}$ which equals 8.

What other math function, F has the properties: $F(9) * F(9) = 9$?

What other math function, G has the property: $G(8) * G(8) * G(8) = 8$?

If we rewrite these two equalities with exponents, we have: $[F(9)]^2 = 9$ and $[G(8)]^3 = 8$. What's going on here?

Hint: Replacing the 9 with 25 would suggest $[F(25)]^2 = 25$

Replacing the 9 with 36 would suggest $[F(36)]^2 = 36$

Replacing the 9 with 100 would suggest $[F(100)]^2 = 100$

Radicals And Rational Expressions

Notice that 9, 25, 36 and 100 are perfect squares

$(9 = 3^2, 25 = 5^2, 36 = 6^2$ and $100 = 10^2)$.

These results suggest that the function F(x) acts like the square root function, i.e. For $x \geq 0$, $F(x) = \sqrt{x}$. By definition, $\sqrt{x} * \sqrt{x} = x$, so $(\sqrt{x})^2 = x$.

Replace the 8 with 27 would suggest $[G(27)]^3 = 27$

Replace the 8 with 125 would suggest $[G(125)]^3 = 125$

Replace the 8 with 1000 would suggest $[G(1000)]^3 = 1000$

Notice that 8, 27, 125 and 1000 are perfect cubes squares

$(8 = 2^3, 27 = 3^3, 125 = 5^3$ and $1000 = 10^3)$

The function G(x) acts like the cube root function,

i.e. For all x, $G(x) = \sqrt[3]{x}$. By definition, $\sqrt[3]{x} * \sqrt[3]{x} * \sqrt[3]{x} = x$, so $(\sqrt[3]{x})^3 = x$.

There are other root functions such as $\sqrt[4]{x}, \sqrt[5]{x}$, etc.

Consider the following: $8^{(\frac{4}{3})} = \left(8^{\frac{1}{3}}\right)^4 = (\sqrt[3]{8})^4 = 2^4 = 16$

Also $8^{(\frac{4}{3})} = \sqrt[3]{8^4} = \sqrt[3]{4096} = 16$

In general we have the following relationship between radicals and rational exponents: *For* $x > 0, x^{(\frac{p}{q})} = \sqrt[q]{x^p} = (\sqrt[q]{x})^p$.

...

19

Practice Problems:

Solve each equation for the variable x.

17. $x^{\frac{-2}{3}} = 4$

18. $(x+3)^{\frac{1}{2}} = \left(\frac{1}{8}\right)^{\frac{1}{3}}$

EXPONENTIAL FUNCTIONS

Consider the following problem: *You start a savings account with $200 in a bank that pays 5% annual interest, compounded yearly. You make no other deposits. How long will it take for the account to be worth $400?*

At the end of the first year the account will be worth $200 + .05*$200 = 1.05*$200 = $210. At the end of the second year the account will be worth $210 + .05*$210 = 1.05*$210 = $220.5. Another way of calculating this last value is 1.05*1.05*200 is which can be expressed as $1.05^2 * 200$. So, at the end of the third year the account will be worth $200*1.05^3$.

Using a graphing calculator with the two equations $y = 200 * 1.05^x$ and $y = 400$ and the intersect function reveals that the two graphs intersect at $(14.21\ years, \$400)$, so the account will be worth $400 in about fourteen years and two and half months.

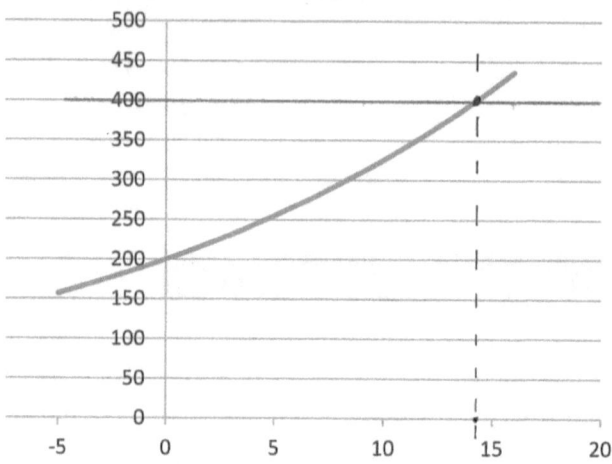

The function $y = 200 * 1.05^x$ is an example of an **exponential function**. The general form of exponential functions is $y = a * b^x$, where x is a variable and a and b are positive constants. The domain of an exponential function is all numbers while the range is all positive numbers.

If the $200 earned 5% interest **compounded monthly**, the equation would be $y = 200(1 + \frac{.05}{12})^{12x}$ where the variable x is in years. In 14 years, the account would be worth $200(1 + \frac{.05}{12})^{(12*14)} = \$402.17$.

If the account earned 5% interest **compounded daily**, the equation would be $y = 200(1 + \frac{.05}{365})^{365x}$ where the variable x is in years. The account would be worth $200(1 + \frac{.05}{365})^{(365*14)} = \$402.73$ in 14 years.

The equation for compounding continuously uses Euler's e as the base. See Appendix B for a discussion regarding Euler's e. The equation would be $y = 200 * e^{.05x}$ and the account would be worth $200 * e^{0.05*14} = \$402.75$.

The **doubling time** of money earning 5% compounded continuously is found by solving the equation $400 = 200 e^{0.05x}$ for the variable x in

Exponential Functions

years. The solution is approximately 13.863 years.

A different equation for the value of the account is $y = 200(2)^{\frac{x}{13.863}}$, where x is in years. Calculating y(14) yields $402.75.

The point of showing both equations $y = 200 * e^{.05x}$ and $y = 200(2)^{\frac{x}{13.863}}$ is to emphasize that the base of an exponential equation is not unique.

To solve the equation $8 = 3^x$ we use a graphing calculator, graphing both the equation $y = 8$ and the equation $y = 3^x$. These graphs intersect at (1.893, 8.002) so the answer is x is approximately 1.893.

Radioactive decay is an example of **continuous declination**. The isotope polonium has a half-life of 138 days and is decaying per the equation, $y = A_0(\frac{1}{2})^{\frac{x}{138}}$, where A_0 is the initial amount of polonium and x is the number of days of decay. The variable y represents the amount remaining after d days.

Note that the base $\frac{1}{2}$ is less than 1 which indicates the amount is declining. For example, after 276 days, $A = G(\frac{1}{2})^{\frac{276}{138}} = G(\frac{1}{2})^2 = \frac{1}{4}G$.

Using a graphing calculator to solve the equation $\frac{1}{2} = e^x$ for x yields an approximate answer $x = -.693$.

A base e exponential equation for the decaying radioactivity of polonium is $= e^{\frac{-.693x}{138}}$. Calculating $y(276) = 0.25007$ which is consistent with the $\frac{1}{4}G$ found using the half-life equation.

...

It is important to notice that in an exponential equation, the **independent variable x is the exponent**. To emphasize this, note that for all x $x^2 \neq 2^x$.

Using algebra to solve for the exact values of exponential variables requires the concept of the **inverse of a function**.

...

INVERSE FUNCTIONS

The inverse of a collection of points reverses the role of the coordinates. For example the inverse of the set $\{(-3,-27),(-2,-8),(-1,-1),(0,0),(1,1),(2,8),(3,27)\}$ is the set $\{(-27,-3),(-8,-2),(-1,-1),(0,0),(1,1),(8,2),(27,3)\}$.

A function is essentially a collection of points, so we derive the rule of the inverse of a function by reversing the roles of x and y then solve for y in terms of x. For example, to find the inverse of $f(x) = \frac{2}{3}x - 5$, re-write the equation as $y = \frac{2}{3}x - 5$ and reverse the role of x and y for $x = \frac{2}{3}y - 5$. Solving for y in terms of x yields $3x = 2y - 15$ then $2y = 3x + 15$ or $y = \frac{3}{2}x + \frac{15}{2}$.

The inverse of a function f is also denoted by f^{-1}, where the superscript -1 is not an exponent, but just another use of -1. So for $f(x) = \frac{2}{3}x - 5$, $f^{-1}(x) = \frac{3}{2}x + \frac{15}{2}$.

To confirm that f and f^{-1} reverses x's and y's we calculate:

$f(-3) = \frac{2}{3}(-3) - 5 = -7$, while $f^{-1}(-7) = \frac{3}{2} * -7 + \frac{15}{2} = \frac{-6}{2} = -3$.
$f(9) = \frac{2}{3}(9) - 5 = 1$, while $f^{-1}(1) = \frac{3}{2} * 1 + \frac{15}{2} = \frac{18}{2} = 9$.

For the notation f^{-1} behaves like an exponent for only the function, $f(x) = x^{-1} = \frac{1}{x}$. So for $f(x) = \frac{1}{x}$, $f^{-1}(x) = \frac{1}{f(x)}$.

By definition functions f and g are inverses if $g(f(x)) = f(g(x)) = x$. Note that inverse functions come in pairs; each is the inverse of the other.

The inverse of the function $f(x) = x^3$ is the function $g(x) = \sqrt[3]{x}$ where $g(f(x)) = g(x^3) = \sqrt[3]{x^3} = x$ and $f(g(x)) = f\left(\sqrt[3]{x}\right) = \left(\sqrt[3]{x}\right)^3 = x$.

...

Practice Problems:

Find the inverse of each function.

19. $f(x) = \dfrac{2}{3}x + \dfrac{4}{5}$

20. $f(x) = \dfrac{1}{x-1}$

...

INVERSE OF EXPONENTIAL FUNCTIONS

With $f(x) = 2^x$, $f(3) = 2^3 = 8$, then $f^{-1}(8)$ equals '*the power of 2 that yields 8*. We abbreviate the phrase '*the power of 2 that yields 8*' with the words '*log-base-2 of 8*' and use the mathematical notation $log_2(8)$. The expression '*log*' stands for the word **logarithm**.

For $f(x) = 2^x$, $f^{-1}(8) = log_2(8) = 3$ because $2^3 = 8$.

For $f(x) = 2^x$, $f^{-1}(4) = log_2(4) = 2$, because $2^2 = 4$.

For $f(x) = 2^x$, $f^{-1}(0) = log_2(0) = 1$, because $2^0 = 1$

For $f(x) = 3^x$, $f^{-1}(81) = log_3(81) = 4$, because $3^4 = 81$

For $f(x) = 10^x$, $f^{-1}(1000) = log_{10}(1000) = 3$, because $10^3 = 1000$.

To calculate $log_{16}(8)$, convert the logarithmic statement $log_{16}(8) = x$ to its equivalent exponential statement $8 = 16^x$.

Then express **16 as 2^4** and **8 as 2^3**. The equation $8 = 16^x$ becomes $2^3 = (2^4)^x$.

From the properties of exponents to $(2^4)^x = (2^x)^4$. We then have the equation $2^3 = (2^4)^x = (2^x)^4$. Raising both sides of the equation $2^3 = (2^x)^4$ to the $\frac{1}{4}$th power yields $(2^3)^{\frac{1}{4}} = (2^x)^{4*\frac{1}{4}} = 2^x$.

From $2^x = 2^{\frac{3}{4}}$, $x = \frac{3}{4}$.

Therefore $log_{16}(8) = \frac{3}{4}$ because $16^{\frac{3}{4}} = (\sqrt[4]{16})^3 = 2^3 = 8$.

Each property of exponents has a corresponding property of logarithms.

For $B > 0$, $B^M * B^N = B^{M+N}$ -> $log_B(M * N) = log_B(M) + log_B(N)$

For $B > 0$, $\frac{B^M}{B^N} = B^{M-N}$ -> $log_B\left(\frac{M}{N}\right) = log_B(M) - log_B(N)$

For $B > 0$, $B^0 = 1$ -> $log_B(1) = 0$

For $B > 0$, $B^{-1} = \frac{1}{B}$ -> $log_B\left(\frac{1}{B}\right) = -1$

For $B > 0$, $B^{-N} = \frac{1}{B^N}$ -> $log_B\left(\frac{1}{B^N}\right) = -N$

For $B > 0$, $(B^M)^N = B^{(M*N)}$ -> $log_B(M^N) = N log_B(M)$

By definition: For $B > 0$, $log_B(A) = E$ if and only if $B^E = A$.

...

Practice Problems:

Solve each equation for the variable x.

21. $log_x\left(\frac{1}{4}\right) = \frac{-1}{2}$

Inverse Of Exponential Functions

22. $log_3(x) = -2$

23. $log_{\frac{1}{3}}(27) = x$

24. $log_7(x+1) + log_7(x-5) = 1$

25. $log(x-3) - log(x-1) = 1$

$\ldots$

Calculators provide values of logarithms of base 10 with the key *log* and logarithms base *e* with the key *ln*. To calculate logarithms with a base other than e or 10, we can use the **change of base** property:

$$log_B(A) = \frac{log_C(A)}{log_C(B)}.$$

For example $log_{16}(8) = \frac{log_2(8)}{log_2(16)} = \frac{3}{4}.$

Most logarithm values are decimal numbers. For example, entering $\frac{log(8)}{log(16)}$ into my calculator provides 0.75.

An important note: Base 10 logarithms are expressed as $log(x)$ and base *e* logarithms are expressed as $ln(x)$.

We return to a previously posed problem: *You start a savings account with $200 in a bank that pays 5% annual interest, compounded yearly. You make no other deposits. How long will it take for the account to be worth $400?*

As discussed earlier, we need to solve the equation $400 = 200 * 1.05^x$ for x. We first reduce the equation to $2 = 1.05^x$ and convert this equation to its corresponding logarithm equation: $x = log_{1.05}(2) = \frac{log(2)}{log(1.05)}.$ My calculator provides 14.20669908

This compares favorably with the answer we found earlier using the graphing calculator, with the two equations $y = 200 * 1.05^x$ and $y = 400$. The value of the account will be $400 in about fourteen years and two and half months.

...

Practice Problems:

26. Determine how much time is required for a $500 investment to double in value if interest is earned at the rate of 4.75% compounded annually.

Smith Hauling purchased an 18-wheel truck for $100,000. The truck depreciates at the constant rate of $10,000 per year for 10 years.

27. Write an expression that gives the value *y* after *x* years.

28. When is the value of the truck $55,000?

...

Consider this problem from calculus. *As a result of Newton's law of cooling, a hot cup of coffee in a room at* 70°F *will cool at a* **rate proportional to the difference** *between the coffee temperature and the room temperature. If the cup of coffee is poured at* 190°F, *has cooled to* 178°F *after one minute, what is the temperature of the coffee after 10 minutes?*

To solve this problem, students first use calculus skills to determine that the equation will have the form $T(t) = 70 + e^C e^{kt}$, where k is the **constant of proportionality**, T is the temperature in Fahrenheit at time t in minutes and C is a number to be determined from the

Inverse Of Exponential Functions

given two temperatures.

From T(0)=190 we have the equation $190 = 70 + e^C e^0$ so $e^C = 120$.

From T(1)=178 we have the equation $178 = 70 + 120e^k$ so $e^k = \frac{178-70}{120} = \frac{108}{120} = .9$ therefore $k = \ln(.9) = -0.1$.

So the equation is $T(t) = 70 + 120e^{-0.1t}$. In ten minutes $T(10) = 70 + 120e^{-0.1*10} = 70 + \frac{120}{e} = 114°F$.

. . .

Practice Problems:

A drug is administered intravenously for pain. The function $f(t) = 90 - 52\ln(1 + t)$ gives the number of units of the drug remaining in the body after t hours.

29. What was the initial number of units of the drug administered?

30. How much is present after 2 hours?

EXPONENTIAL AND LOGARITHMIC GRAPHS

The figure below shows the graphs of the **increasing function** $y = e^{1.5x}$ with the base $e^{1.5} > 1$ and the **decreasing function** $y = e^{-1.2x}$ with the base $\frac{1}{e^{1.2}} < 1$. Both functions are positive for all values of x, therefore both their graphs are above the x-axis.

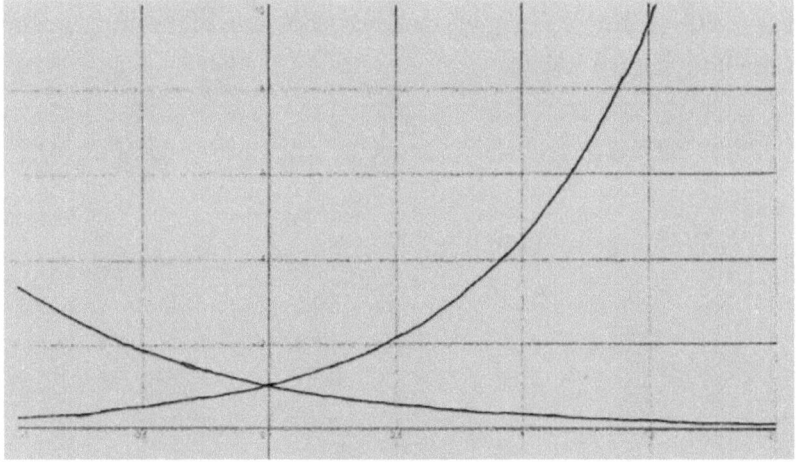

When the base of an exponential function is larger than one, it is sometimes referred to as **exponential growth.** Populations often experience exponential growth over a finite amount of time. When the base of an exponential function is larger than zero and less than one, it is sometimes referred to as **exponential decay.** Radioactive isotopes decay exponentially.

Practice Problems:

31. The population of Glenbrook is 375,000 and is increasing at the rate of 2.25% per year. Predict when the population will be 1 million.

Suppose that a colony of bacteria starts with 1 bacterium and doubles in number every half hour.

32. How many bacteria will the colony contain at the end of 2 hours?

33. How many bacteria will the colony contain at the end of 4 hours?

34. How many bacteria will the colony contain at the end of 24 hours?

The half-life of a certain radioactive substance is 12 hours. There are 8 grams present initially.

35. Express the amount of substance remaining as a function of time t.

36. When will there be 1 gram remaining?

37. Suppose you desire to double your $500 investment in ten years. What interest rate, compounded annually, would be required?

38. The decay equation for radon-222 gas is known to be

$y = y_0 e^{-0.18t}$ with t in days. About how long will it take the radon in a sealed sample of air to fall to 90% of its original value?

The figure below shows the graphs of the increasing function $y = 2^x$, the increasing function $y = \log_2(x)$ and the line with equation $y = x$. The two functions $y = 2^x$ and $y = \log_2(x)$ are inverses of each other. The graphs of a function and its inverse are symmetric with respect to the line $y = x$.

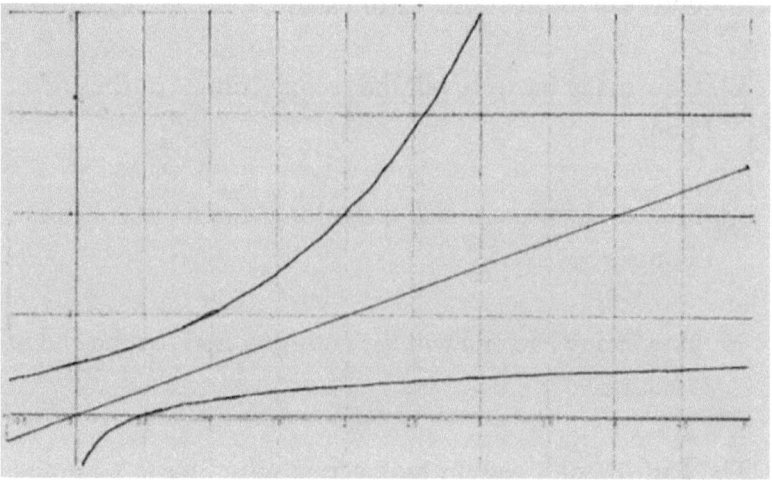

All exponential functions, such as $f(x) = 2^x$ are positive for every value of x. All logarithmic functions, such as of $f(x) = \log_2(x)$ are restricted to positive values of x. Division by zero is not permitted and even roots of negative numbers are not permitted. A third restriction is a logarithm of zero and a logarithm of a negative number.

A logarithm can be zero ($\log_2(1) = 0$) and a logarithm can be a negative number ($\log_2\left(\frac{1}{4}\right) = -2$, but an expression containing $\log_B(negative)$ is never permitted. It is best to use absolute value notation such as $\log_B(|expression|)$ when the expression contains a variable.

RATIONAL ALGEBRAIC FUNCTIONS

Consider the following problem: *A large pump can empty the town swimming pool in 7 hours less than a smaller one. Together they can empty the pool in 12 hours. How long would it take the large pump alone to empty the pool? How long would it take the small pump alone to empty the pool?*

Let x represent the number of hours it would it take the large pump, alone to empty the pool, then this large pump empties $\frac{1}{x}$ of the pool each hour. For example, if the large pump can empty the tank in 3 hours, this pump empties $\frac{1}{3}$ of the pool each hour.

The small pump empties $\frac{1}{x+7}$ of the pool each hour. In 12 hours the large pump will empty $\frac{12}{x}$ part of the pool and the small pump will empty $\frac{12}{x+7}$ part of the pool.

In 12 hours the whole pool is emptied, so we have the equation $\frac{12}{x} + \frac{12}{x+7} = 1$. This equation can be solved by multiplying both sides of the equation by $x(x+7)$ resulting in the equation $12(x+7) + 12x = x(x+7)$. We then have $24x + 84 = x^2 + 7x$ or $0 = x^2 - 21x - 84 = (x-21)(x+4)$.

Working alone, the large pump can empty the pool in 21 hours and the small would require an additional 7 hours (in 28 hours).

The function $f(x) = \frac{12}{x} + \frac{12}{x+7}$ is an example of a **rational algebraic**

function. Such functions will often have restricted domains because denominators cannot be equal to zero.

One type of rational algebraic functions is the reciprocal of polynomials. The simplest of these are reciprocals of linear and quadratic functions. Some examples are:

a. $y = \dfrac{1}{x}$

b. $y = \dfrac{1}{x-1}$

c. $y = \dfrac{1}{2x+3}$

d. $y = \dfrac{1}{x-\frac{4}{7}} = \dfrac{7}{7x-4}$

e. $y = \dfrac{1}{\frac{3}{2}x+\frac{2}{5}} = \dfrac{10}{15x+6}$

f. $y = \dfrac{1}{(x+3)(2x+5)}$

g. $y = \dfrac{-2}{(x-4)(2x+7)}$

Each of these functions has at least one **vertical asymptote** occurring where the x values yields a denominator equal to zero.

For a. this is at $x = 0$,

for b. this at $x = 1$

for c. at $x = \dfrac{-3}{2}$

for d. $x = \dfrac{4}{7}$

for e. at $x = \dfrac{-2}{5}$

for f. at $x = -3$ and $x = \frac{-5}{2}$

for g. at $x = 4$ and $x = \frac{-7}{2}$

The graph of $y = \frac{-2}{(x-4)(2x+7)}$ is shown below. The graph has **discontinuities** at $x = 4$ and $x = \frac{-7}{2}$.

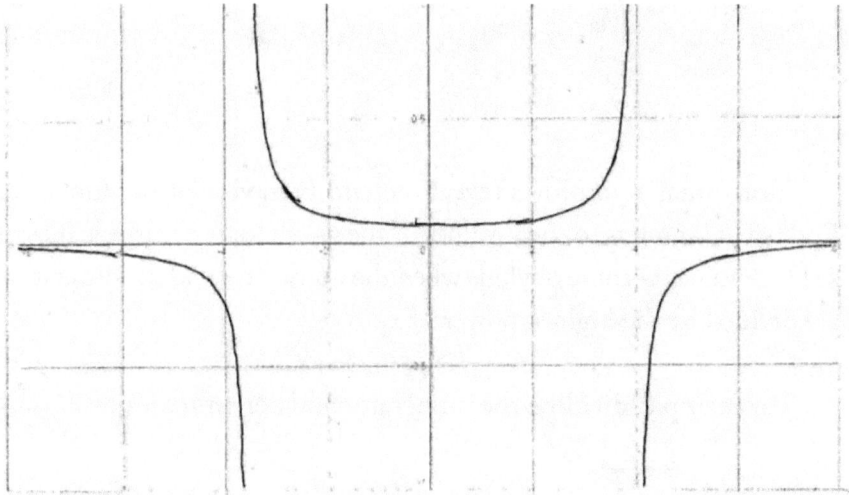

The function $y = \frac{3x^2+2}{x^2+4}$ does not have a vertical asymptote because $x^2 + 4$ is positive for all values of x. The function's graph below shows that the **horizontal asymptote is $y = 3$**.

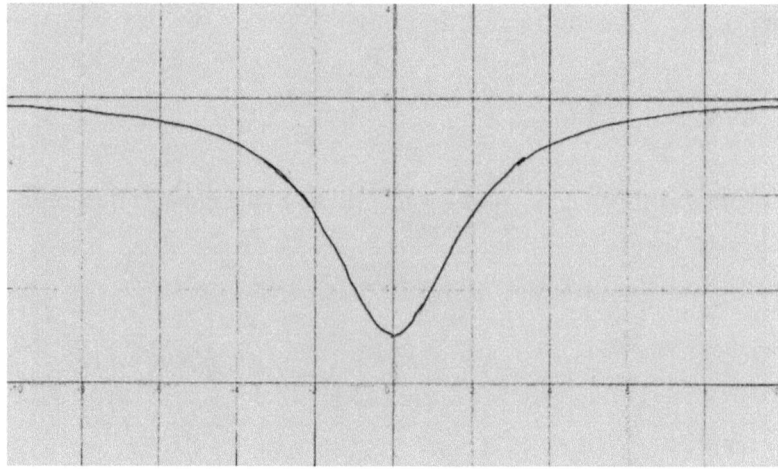

Horizontal asymptotes reveal the **end behavior** of the function. End behavior refers to the y-values of the graph for very large values of x. These values become obvious when the numerator and denominators are divided by the highest power of x.

For example, dividing the numerator and denominator of $\frac{3x^2+2}{x^2+4}$ by x^2 yields $\frac{3+\frac{2}{x^2}}{1+\frac{4}{x^2}}$.

For very large values of x, the expressions $\frac{2}{x^2}$ and $\frac{4}{x^2}$ are very small and essentially zero, so consider the expressions $\frac{3+\frac{2}{x^2}}{1+\frac{4}{x^2}}$ to be $\frac{3+0}{1+0}$ or $\frac{3}{1}$.

The equation of the horizontal asymptote is $y = 3$.

Some problems in calculus ask for both vertical and horizontal asymptotes. A function can have several vertical asymptotes. But it can have at most two horizontal asymptotes, one positive and one negative.

Consider the function $y = \frac{10x+2}{\sqrt{x^2+4}} = \frac{10+\frac{2}{x}}{\sqrt{(\frac{x^2}{x^2}+\frac{4}{x^2})}} = \frac{10+\frac{2}{x}}{\sqrt{(1+\frac{4}{x^2})}}$.

Rational Algebraic Functions

The graph of this function below reveals has two horizontal asymptotes $y = 10$, and, $y = -10$.

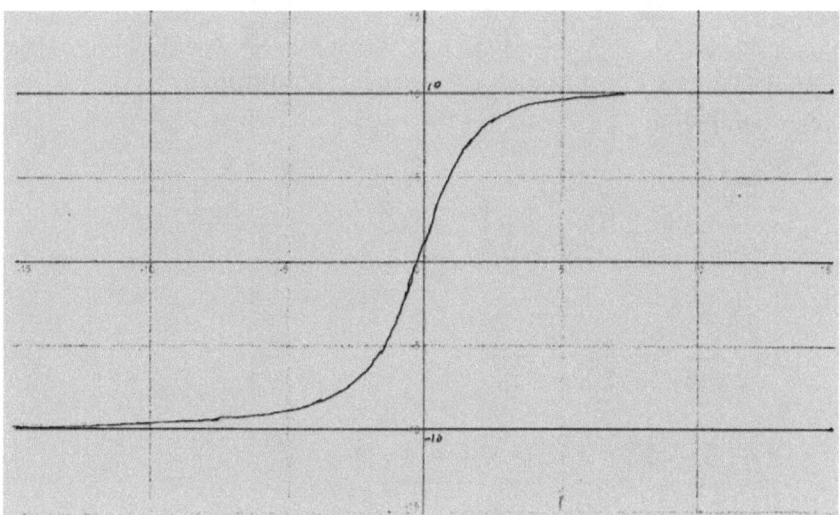

Practice Problems:

For each function determine the equations the vertical asymptotes.

39. $y = \dfrac{x^4+1}{x^4-5x^2+4}$

40. $y = \dfrac{1}{7x^3-5x^2+7x-5}$

For each function determine the equation of the horizontal asymptotes.

41. $y = \dfrac{-4x^2+1}{x^2+25}$

42. $y = \dfrac{-5x+1}{\sqrt{x^2+4}}$

In calculus, there will be problems requiring transforming a **bi-linear expression** $\dfrac{ax+b}{cx+d}$ to the reciprocal of a linear expression plus a constant

For example, the expression $\dfrac{x+2}{x+1}$ canl be transformed to $\dfrac{1}{x+1}+1$.

Changing these bi-linear forms can often require several transformations. For example, $\dfrac{2x}{3x-1}$ can be transformed to $\dfrac{2}{3}\left(\dfrac{-1}{3x+1}+1\right)$ as shown below.

$$\dfrac{2x}{3x-1} = 2\left(\dfrac{x}{3x+1}\right)$$

$$2\left(\dfrac{x}{3x+1}\right) = \dfrac{2}{3}\left(\dfrac{3x}{3x+1}\right)$$

$$\dfrac{2}{3}\left(\dfrac{3x}{3x+1}\right) = \dfrac{2}{3}\left(\dfrac{3x+1-1}{3x+1}\right)$$

$$\dfrac{2}{3}\left(\dfrac{3x+1-1}{3x+1}\right) = \dfrac{2}{3}\left(\dfrac{3x+1}{3x+1} - \dfrac{1}{3x+1}\right)$$

$$\dfrac{2}{3}\left(\dfrac{3x+1}{3x+1} - \dfrac{1}{3x+1}\right) = \dfrac{2}{3}\left(\dfrac{-1}{3x+1} + 1\right)$$

In calculus, there will also be problems requiring transforming the reciprocal of a quadratic expression to the sum of the reciprocals of two linear expressions. For example, given $y = \dfrac{1}{(2x-1)(x+4)} = \dfrac{A}{2x-1} + \dfrac{B}{x+4}$ the requirement is to determine the value of A and B. The details are shown below.

$$\dfrac{A}{2x-1} + \dfrac{B}{x+4} = \dfrac{A(x+4)}{(2x-1)(x+4)} + \dfrac{B(2x-1)}{(2x-1)(x+4)} =$$

$$\dfrac{Ax + 4A + 2Bx - B}{(2x-1)(x+4)} = \dfrac{(A+2B)x + (4A-B)}{(2x-1)(x+4)} =$$

Rational Algebraic Functions

$$\frac{1}{(2x-1)(x+4)} = \frac{0x+1}{(2x-1)(x+4)}$$

Therefore $A + 2B = 0$ and $4A - B = 1$. Solving this system of two linear equations yields $A = \frac{2}{9}$ and $B = \frac{-1}{9}$.

So $\frac{1}{(2x-1)(x+4)} = \frac{2}{9}\left(\frac{1}{2x-1}\right) - \frac{1}{9}\left(\frac{1}{x+4}\right)$

Practice Problems:

Transform each expression to an appropriate expression per the discussions above.

43. $\frac{x-3}{x+2}$

44. $\frac{2x-1}{3x+2}$

45. Given $y = \frac{1}{(2x-1)(x+4)} = \frac{A}{2x-1} + \frac{B}{x+4}$ find the values of A and B.

TRIGONOMETRIC FUNCTIONS

The function $f(x) = 37 \sin\left(\frac{2\pi}{365}(x - 101)\right) + 25$ approximates the average Fahrenheit temperature in Fairbanks, Alaska, on day x, during typical 365-day year. Its graph, shown below, is a **sine wave,** sometimes referred to as a **sinusoid.** The function is **periodic,** meaning the graph repeats the behavior after one **period.** The period of the function is 365 days. The variable x is in **radians** as opposed to **degrees.**

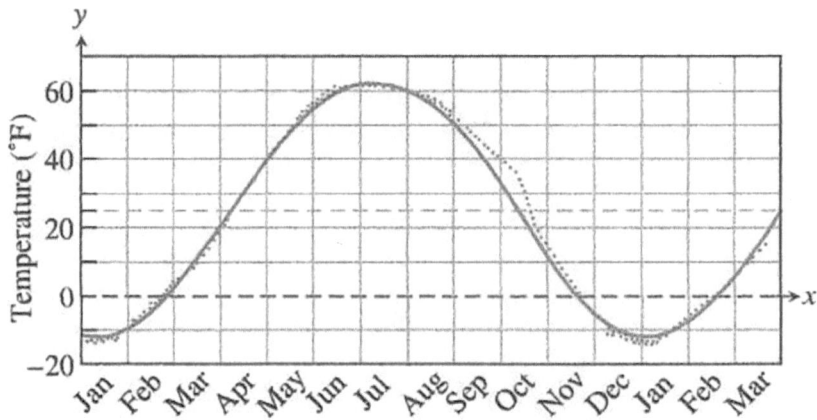

The function is an example of a **trigonometric function.** The **amplitude** of a sine wave is half the distance from the minimum to the maximum of the function, or $\frac{maximum - minimum}{2}$. The amplitude of the function is $\frac{62-(-12)}{2} = 37$.

Trigonometric Functions

The standard form of a trigonometric function is $f(x) = A\sin(Bx + C) + D$.

The **amplitude** is $|A|$, the **period** is $\frac{2\pi}{|B|}$ and the value of $|\frac{C}{B}|$ determines the amount the basic **sine function**, $f(x) = \sin(x)$, is shifted.

The unit-circle has equation, $x^2 + y^2 = 1$, where $y = \sin(\theta)$ and $x = \cos(\theta)$, and θ is the angle determined by the line through the origin and the point with coordinates (x, y).

In the drawing below the distances OC and OS equal 1, the variable θ is in degrees, while the variable x is in radians. The distance OE equals sec(x). The drawing reveals two important identities: $\sin^2(x) + \cos^2(x) = 1$ and $1 + \tan^2(x) = \sec^2(x)$.

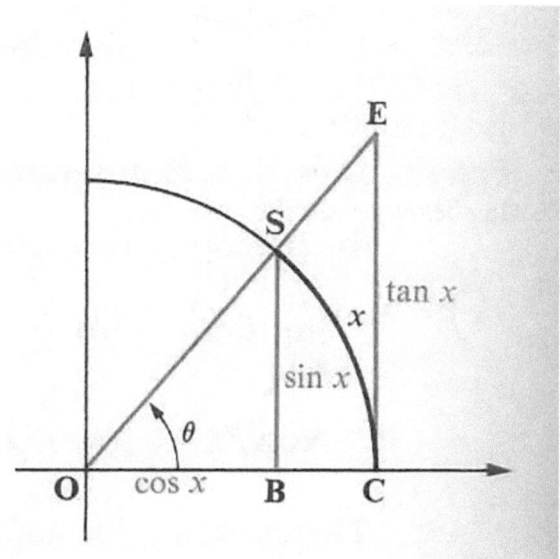

Arcs on the unit circle have the same radian measure as the corresponding angle. One radian equals $\frac{180}{\pi} = \approx 57.30 \; degrees$.

The drawing below reveals several identities:

$$\sin(x) = \sin(\pi - x) = -\sin(\pi + x) = \sin(2\pi - x)$$

$$\cos(x) = \cos(2\pi - x) = -\cos(\pi + x) = -\cos(\pi - x)$$

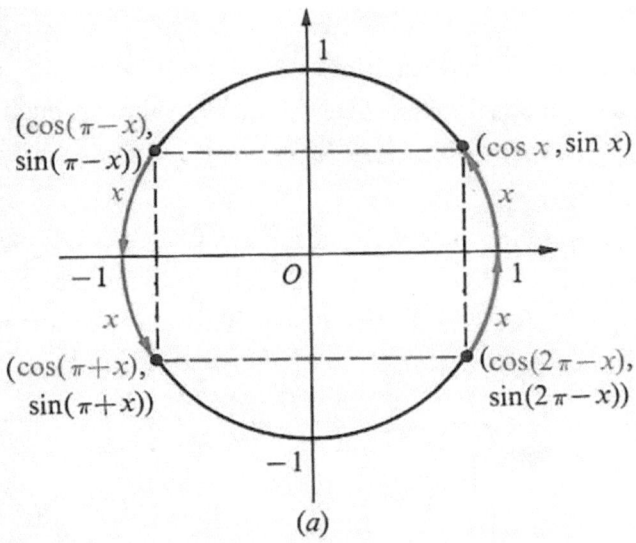

(a)

The drawing below reveals the identities:

$$\cos\left(\frac{\pi}{2} - x\right) = -\cos\left(\frac{\pi}{2} + x\right) = sin(x)$$

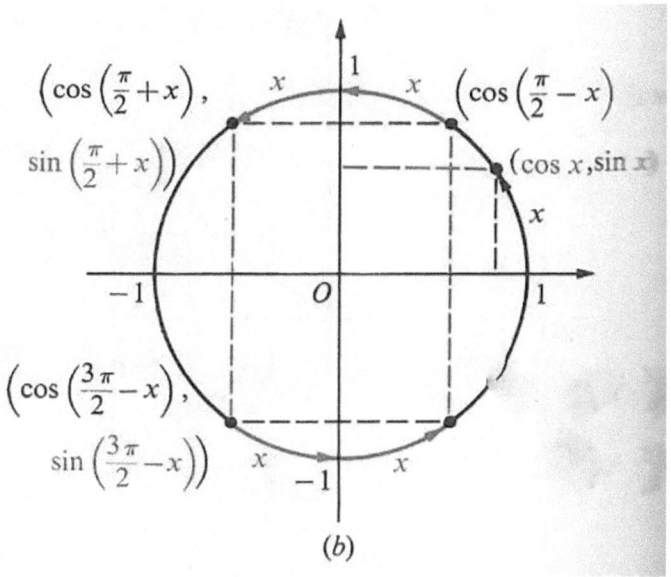

(b)

The drawing below reveals the identities:

$$\tan(x) = \tan(\pi - x) = -\tan(\pi + x) = -\tan(2\pi - x)$$

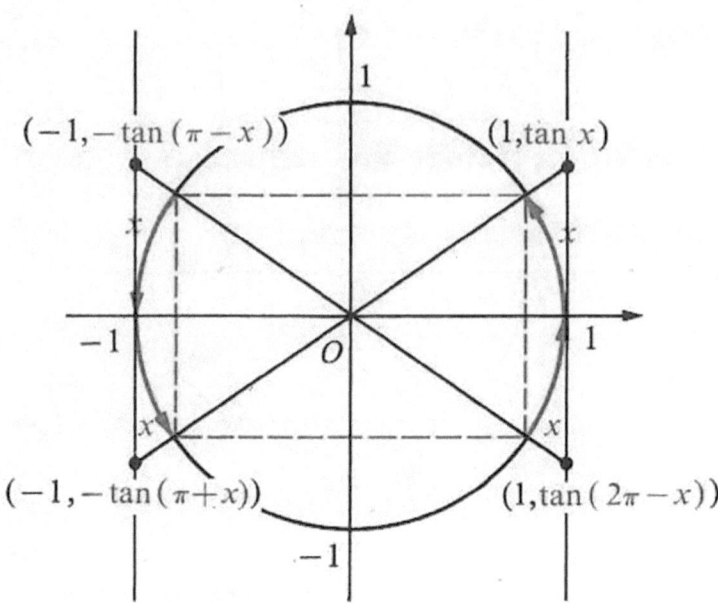

Students of calculus should be familiar with how radian measures correspond to degree measure. For example radian measures $\frac{\pi}{2} = 90°, \pi = 180°, \frac{3\pi}{2} = 270°$ and $2\pi = 360°$. They should also memorize the following sine and cosine values:

$$\sin(30°) = \sin\left(\frac{\pi}{6}\right) = \frac{1}{2} = \cos(60°) = \cos\left(\frac{\pi}{3}\right)$$

$$\sin(60°) = \sin\left(\frac{\pi}{3}\right) = \frac{\sqrt{3}}{2} = \cos(30°) = \cos\left(\frac{\pi}{6}\right)$$

$$\sin(45°) = \sin\left(\frac{\pi}{4}\right) = \frac{\sqrt{2}}{2} = \cos(45°) = \cos\left(\frac{\pi}{4}\right)$$

...

We modify a problem presented previously: A rock is thrown

upward at an angle of 35° and 48 feet off of the roof of a building with a velocity of 60 mph. The rock will travels in an arc, first moving upward, then continuing downward to the ground. A mathematical model for this situation is the two parametric equations $x = 88\cos(35°)t$ and $y = 48 + 88\sin(35°)t - 16t^2$. The graph below shows the actual flight path of the rock.

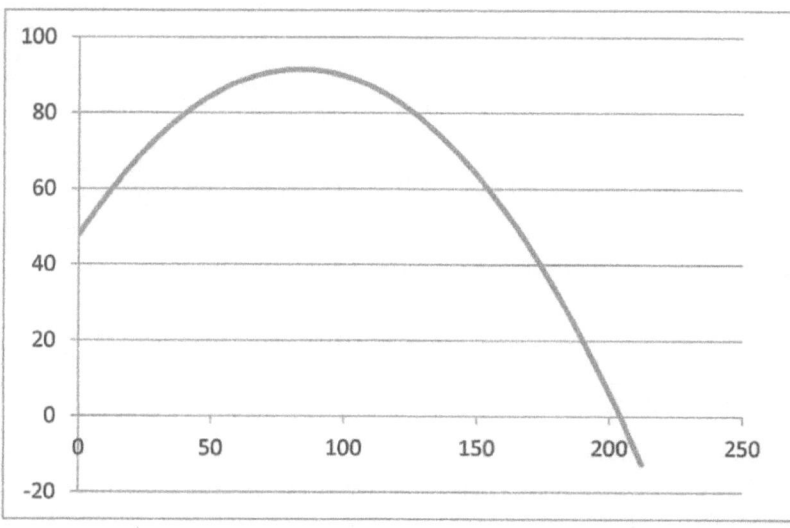

The rock hits the ground a little over 200 feet after 4 seconds of travel.

APPENDIX A

The Number Zero

While tutoring elementary school students, I have often been asked question about the number zero. For example, Lisa, a fifth grade student asked me to explain why division by zero is wrong. I then asked her, 'What you want five divided zero to equal?'

Lisa responded, "Zero."

I then wrote, $\frac{5}{0} = 0$ is false because it does not agree with multiplication. I told Lisa, "Division is justified by multiplication. For example, $\frac{12}{3} = 4$ because $3 * 4 = 12$."

Lisa thought for a moment and seemingly understood the explanation because she then stated, "So, $\frac{0}{0} = 1$, because $1 * 0 = 0$. Is that correct?"

"Lisa, your multiplication statement is correct, but notice that $2 * 0 = 0$, and $5 * 0 = 0$, and $12 * 0 = 0$. In fact any number times zero equals zero.

If a mathematician was building a calculator, she would recognize that any number was possible. Mathematicians want only one answer for an arithmetic calculation, so they decided not to provide an answer.

Appendix A

When either $\frac{5}{0}$ or $\frac{0}{0}$ is entered on a calculators, the device will display 'ERROR" or "DIVISION BY ZERO'."

In calculus, you will encounter situations where A VERY SMALL NUMBER DIVIDED BY A VERY SMALL NUMBER EQUALS A NUMBER.

APPENDIX B

Dealing With Numerical Fractions

There is an easy technique used to eliminate numerical fractions and fractional coefficients in equations. The technique uses an important observation from arithmetic, namely $x * \frac{y}{x} = x$ and $\frac{y}{x} * x = x$. Some numerical exanmples are: $5 * \frac{3}{5} = 3, 12 * \frac{9}{12} = 9$ and $\frac{33}{125} * 125 = 33$.

Consider the equation $\frac{1}{2}x - 3 = 7 + \frac{5}{2}x$. Multiplying both sides of the equation by 2, will transform the equation to the equivalent equation: $x - 6 = 14 + 5x$. We then have $-20 = 4x$. So $x = -5$.

Consider the quadratic equation $x^2 + \frac{5}{6}x - \frac{2}{3} = 0$. Multiplying both sides by 6 results in the equation $6x^2 + 5x - 4 = 0$ which can be factored into $(2x - 1)(3x + 4) = 0$, with solution $x = \frac{1}{2}$ and $x = \frac{-4}{3}$.

This technique (multiplying both sides of an equation by a common denominator of all the fractions) eliminates the need to add or subtract these fractions. The final answer might require dividing an integer by an integer.

. . .

When solving equations or inequalities, fewer mistakes are made

Appendix B

by maintaining positive coefficients of the variable in question. For example, given the inequality $17 - 3x \geq 7x + 20$, add $3x$ to both sides of the $\geq$ symbol and subtract 20 from both sides of the $\geq$ symbol to obtain $-3 \geq 10x$.

Divide both sides by 10 for $\dfrac{-3}{10} \geq x$. Therefore $x \leq \dfrac{-3}{10}$.

APPENDIX C

Euler's e

Earlier we used the equation $y = 200(1+\frac{.05}{12})^{12x}$ to explore interest **compounded monthly** and the equation $y = 200(1+\frac{.05}{365})^{365x}$ to explore interest **compounded daily.** In each case the account began with $200 and earned 5% annual interest. There are 8760 hours in a year, so we could use the equation $y = 200(1+\frac{.05}{8760})^{8760x}$ to explore interest **compounded hourly.** We could continue on to compounding every minute then to compounding every second. If the rate was 100% we would explore the equation $y = 200(1+\frac{1}{N})^{Nx}$ where N is very large. It out that the expression $(1+\frac{1}{N})^N$ becomes arbitrarily close to specific number named in honor of the mathematician Leon Euler. This number is approximately 2.718281828 and is called e. So our equation becomes $y = 200 * e^x$ for a 100% interest rate and $y = 200 * e^{.05x}$ for a 5% interest rate.

The figure below shows graphs of the three functions $y = 3^x$ increasing the fastest followed by $y = e^x$ and $y = 2^x$.

Appendix C

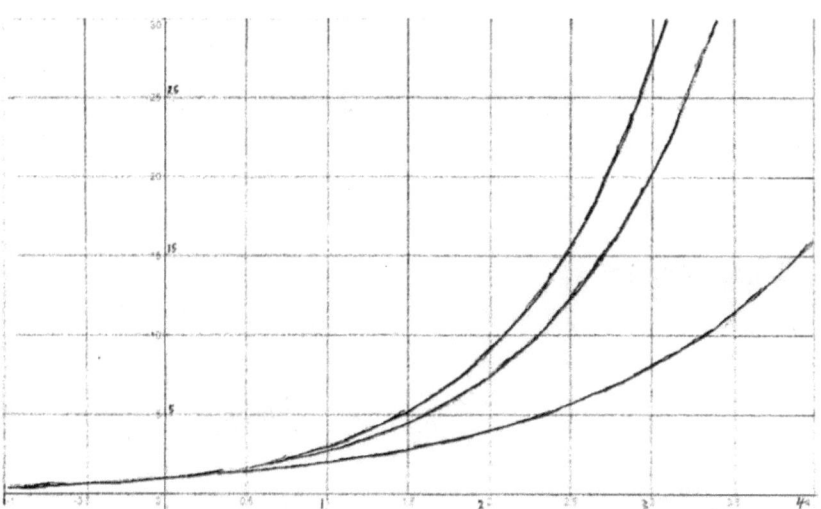

APPENDIX D

Answers Practice Problems

Use a graphing calculator to express each polynomial as a product of linear and quadratic factors.

1. $6x^5 + 19x^4 - 5x^3 - 25x^2 - x + 6.$

The graph of $y = 6x^5 + 19x^4 - 5x^3 - 25x^2 - x + 6$

reveals zeros at $x = 13, x = -1, x = \frac{-2}{3}, x = \frac{1}{2},$ and $x = 1.$

The polynomial can be expressed as

$(x + 3)((x + 1)(3x + 2)(2x - 1)(x - 1)$

2. $3x^4 + x^3 - 4x^2 - 7x - 5.$

The graph of $y = 3x^4 + x^3 - 4x^2 - 7x - 5$

reveals zeros at $x = -1$ and $x = \frac{5}{3}$

so the linear factors are $(x + 1)(3x - 5) = 3x^2 - 2x - 5.$

Dividing By dividing $3x^4 + x^3 - 4x^2 - 7x - 5$ by $3x^2 - 2x - 5$ yields

Appendix D

the prime quadratic $x^2 + x + 1$. The polynomial can be expressed as $(x + 1)(3x - 5)(x^2 + x + 1)$.

Three hundred books currently sell for $40 each, resulting in a revenue of (300)($40) = $12,000. Research indicates that for each $5 increase in the price, 25 fewer books are sold.

3. Write the revenue R as a function of the number x of $5 increases.

$R(x) = (300 - 25x)(40 + 5x)$.

4. What selling price will bring the maximum revenue?

The maximum occurs at (2, 12500) so the price should be $40+2*5=$50.

A 24 inch by 36 inch sheet of cardboard is folded in half to form a 24 inch by 18inch rectangle as shown in the figure below. Four congruent squares of side length x are cut from the corners of the folded rectangle. The sheet is unfolded and the six tabs are folded up to form a cardboard suitcase.

5. Write a formula V(x) for the volume of the suitcase. The 36 inch side consist of $x + 2x + 2x + 2x + x = x + 2x + (36 - 6x) + 2x + x$ and the 24 inch side consist of x+(24-2x)+x. So the dimensions of the base of the suitcase are $(24 - 2x)$ by $(36 - 6x)$. A formula for the volume is $V = 2x(24 - 2x)(36 - 6x)$.

6. Find the domain of V. Domain is 0<x<6.

7. Find a value of x that yields a volume of 1120 cubic inches. The equation $2x(24 - 2x)(36 - 6x) = 1120$ has solutions $x = .80\ and\ x = 4.63$.

8. Use a graphing calculator to find the maximum volume and the value of x yielding this volume. The graph of $y = 2x(24 - 2x)(36 - 6x)$ has a maximum at $(2.54, 1995.32)$.

|-x + 2x +   2x +   2x + x-|

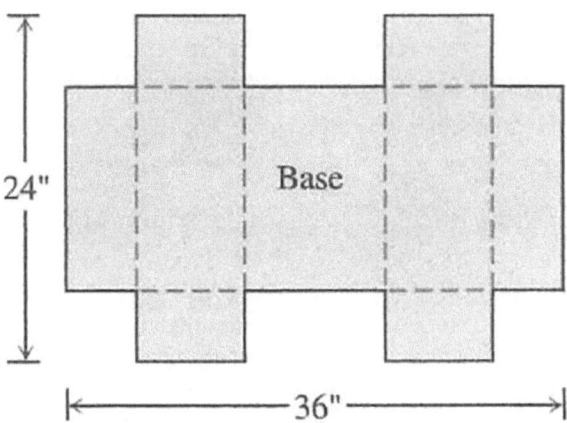

|-x + 2x + (36 - 2x) + 2x + x-|

A rock is thrown upward by an astronaut standing on the moon with a velocity of 100 mph. If the rock leaves the astronaut's hand at 6 feet above the surface of the moon, the mathematical model for this situation is $H = -2.65t^2 + 146.67t + 6$.

The variable H represents the height of the rock in feet above the surface of the moon. The variable t represents the time in seconds. The 6 in the formula is the 6feet from the initial height of the rock. T he -2.65 (feet per second per second) is the effect of gravity on the moon, forcing the rock downward.

9. How much time will it takes for the rock to hit the ground?

Appendix D

Solving the equation $-2.65t^2 + 146.67t + 6 = 0$ for t yields t is approximately 55 seconds.

10. What is the maximum height reached by the rock? The maximum height is approximately 2,017 feet above the surface of the moon.

Express the solution to each inequality as an interval or combination of intervals.:

11. $|4x + 7| \geq 12$

We have $|x + \frac{7}{4}| \geq 3$ so $x + \frac{7}{4} \leq -3$ or $3 \leq x + \frac{7}{4}$

$x \leq -3 - \frac{7}{4}$ or $3 - \frac{7}{4} \leq x$ so $x \leq -\frac{19}{4}$ or $\frac{5}{4} \leq x$

12. $|\frac{2}{3}x - \frac{4}{5}| \leq 6$ Multiply by 15 then we have $|6x - 12| \leq 90$ then divide by 6 for $|x - 2| \leq 15$, so $-15 + 2 \leq x$ and $x \leq 15 + 2$ or $-13 \leq x \leq 17$.

Solve each equation for the variable N.

13. $(2^{3N})^2 = (2^N)^3 * 2^{N+1}$

The equation reduces to $2^{6N} = 2^{3N+N+1}$

so $6N = 4N + 1$, or $N = \frac{1}{2}$

14. $4^{N+3} * 16^N = 8^{3N}$

Changing each term to base-2, the equation transforms to $2^{2N+6} * 2^{4N} = 2^{9N}$, so $6N + 6 = 9N$ or $6 = 3N$. So $N = 2$.

15. $\frac{2^{3N}}{2^N} * \frac{2^N}{2^{5N}} = \frac{4^{-N}}{4^{3N}}$

57

16. *The radius of each of the 79 electrons in a gold atom is approximately 2.80E-13 meters and the radius of the gold atom is 2.88E – 10. Find the proportion of space in the gold atom contained by the electrons.*

To find the proportion of space in the gold atom contained by the 79 electrons, we form the fraction: $\dfrac{79*2.80*10^{-13}}{2.88*10^{-9}}$ and use the properties of exponents to simply the ratio. The ratio equals $\dfrac{227.52*10^{-13}}{2.88*10^{-9}} = \dfrac{2.27*10^{-11}}{2.88*10^{-9}} = \dfrac{2.27}{2.88} * 10^{9-11} = .788 * 10^{-2}$ or approximately $.00788$

The proportion of space in the gol atom contained by the electrons is approximately 0.00788. A gold atom is denser than a carbon atom.

Solve each equation for the variable x.

17. $x^{\frac{-2}{3}} = 4$ Raise both to the $\dfrac{-3}{2}$ power for $x^{\frac{-2}{3}*\frac{-3}{2}} = x$ and

$4^{\frac{-3}{2}} = \dfrac{1}{(\sqrt{4})^3} = \dfrac{1}{2^3} = \dfrac{1}{8}$. Then $x = \dfrac{1}{8}$.

18. $(x+3)^{\frac{1}{2}} = \left(\dfrac{1}{8}\right)^{\frac{1}{3}}$ Raise both to the 6th power for

$(x+3)^3 = \left(\dfrac{1}{8}\right)^2$

then we have $(x+3)^3 = (2^{-3})^2 = (2^2)^{-3}$ or $(x+3)^3 = \dfrac{1^3}{4}$

so $x + 3 = \dfrac{1}{4}$ then $x = \dfrac{-11}{4}$.

Find the inverse of each function.

19. $f(x) = \dfrac{2}{3}x + \dfrac{4}{5}$

Let $y = \dfrac{2}{3}x + \dfrac{4}{5}$ then interchange x and y for

Appendix D

$x = \frac{2}{3}y + \frac{4}{5}$ then $15x = 10y + 12$, so $y = \frac{15}{10}x - \frac{12}{10}$.

So $f^{-1} = \frac{3}{2}x - \frac{6}{5}$.

20. $f(x) = \frac{1}{x-1}$

Let $y = \frac{1}{x-1}$ then interchange x and y for $x = \frac{1}{y-1}$

then $xy - x = 1$ or $xy = x + 1$.

Then $y = \frac{x+1}{x}$ so $f^{-1} = \frac{x+1}{x}$.

Solve each equation for the variable x.

21. $\log_x\left(\frac{1}{4}\right) = \frac{-1}{2}$

In exponential form we have $x^{\frac{-1}{2}} = \frac{1}{4}$.

Raise both to the -2 power for $x = \left(\frac{1}{4}\right)^{-2} = (4)^3 = 16$.

22. $\log_3(x) = -2$

In exponential form we have $x = 3^{-2} = \frac{1}{9}$.

23. $\log_{\frac{1}{3}}(27) = x$

In exponential form we have $\left(\frac{1}{3}\right)^x = 27 = 3^3 = \left(\frac{1}{3}\right)^{-3}$

so $x = -3$.

24. $\log_7(x+1) + \log_7(x-5) = 1$

Combine the two logs for $\log_7[(x+1)(x-5)] = 1$

In exponential form we have

$(x+1)(x-5)] = 7^1 = 7.$

Then $x^2 - 4x - 5 - 7 = 0$. or

$x^2 - 4x - 12 = (x+2)(x-6) = 0$

or $x = -2$ or $x = 6$.

Check by substituting these values in the original and we have

$log_7(-2+1) + log_7(-2-5) = 1$

where log of negative is disallowed

$log_7(6+1) + log_7(6-5) = log_7(7) + log_7(1) = 1 + 0 = 1$

25. $log(x-3) - log(x-1) = 1$

Combine the two logs for $\log\left(\frac{x-3}{x-1}\right) = 1$. So $\frac{x-3}{x-1} = 10^1$

then $x - 3 = 10x - 10$ or $9x = 7$. So $x = \frac{7}{10}$.

26. Determine how much time is required for a $500 investment to double in value if interest is earned at the rate of 4.75% compounded annually.

Smith Hauling purchased an 18-wheel truck for $100,000. The truck depreciates at the constant rate of $10,000 per year for 10 years.

27. Write an expression that gives the value y after x years.

28. When is the value of the truck $55,000?

A drug is administered intravenously for pain. The function $f(t) = 90 - 52\ln(1 + t)$ gives the number of units of the drug remaining in the body after t hours.

29. What was the initial number of units of the drug administered?

30. How much is present after 2 hours?

31. The population of Glenbrook is 375,000 and is increasing at the rate of 2.25% per year. Predict when the population will be 1 million.

Suppose that a colony of bacteria starts with 1 bacterium and doubles in number every half hour.

32. How many bacteria will the colony contain at the end of 2 hours?

33. How many bacteria will the colony contain at the end of 4 hours?

34. How many bacteria will the colony contain at the end of 24 hours?

The half-life of a certain radioactive substance is 12 hours. There are 8 grams present initially.

35. Express the amount of substance remaining as a function of time t.

36. When will there be 1 gram remaining?

37. Suppose you desire to double your $500 investment in ten years. What interest rate, compounded annually, would be required?

38. The decay equation for radon-222 gas is known to be $y = y_0 e^{-0.18t}$ with t in days. About how long will it take the radon in a sealed sample of air to fall to 90% of its original value?

For each function determine the equations the vertical asymptotes.

39. $y = \dfrac{x^4 + 1}{x^4 - 5x^2 + 4}$

40. $y = \dfrac{1}{7x^3 - 5x^2 + 7x - 5}$

For each function determine the equation of the horizontal asymptotes.

41. $y = \dfrac{-4x^2 + 1}{x^2 + 25}$

42. $y = \dfrac{-5x + 1}{\sqrt{x^2 + 4}}$

Transform each expression to an appropriate expression per the discussions above.

43. $\dfrac{x-3}{x+2}$

44. $\dfrac{2x-1}{3x+2}$

www.ingramcontent.com/pod-product-compliance
Lightning Source LLC
LaVergne TN
LVHW041543060526
838200LV00037B/1113